The energy

Europe: Myths and

Truths

Charalampos Aposporis

TO FROSSO, EVGENIA AND

NAUSICAA

Contents

1. Introduction

There are times when a crisis occurs in a sector of the economy and it goes unseen by the wider public, since it does not directly affect the average citizen. It will occupy the minds of people in a certain sector, as well as the ministries and specialists, but not everyone else. There are also crises that cause temporary turbulence, they shake stock markets for a while, but they do not result in a deep change in our lives. Some other times, crises turn into something bigger, threatening everyday life and changing things for good.

We do not yet know in which category the energy crisis of 2021-2022 belongs, because it is not over yet. Perhaps it is its very duration that will judge its effects on our civilization and way of life. Although it is possible to return to normality in terms of energy prices and availability, most experts today agree that this is a strong quake that will bring big changes in the long term.

This crisis is global, multifaceted and looks like a Lernaean Hydra. It concerns all parts of the energy sector and affects everyone equally, producers and consumers. It has a truly global reach, hurting countries big and small, rich and poor. It has economic, geopolitical, technological, environmental and cultural features that cannot be analyzed separately.

However, deep down it is a moral crisis, as we are going to find out, that has to do with what "energy" means to the average person.

As millions of citizens around the planet had to spend more time dealing with something they previously considered a given, or even dull, the public dialogue about energy is now of vital importance.

Respectively, it is important to find out what misconceptions exist in the public discourse for these specialized issues, in order for the citizen to be able to draw conclusions free of fake news, interests, stereotypes and wrongful beliefs.

Since we are now living in such a time, it is hard for someone to penetrate a sector that looks gigantic in scope, but also opaque, such as energy. Indeed, the energy sector is an entire universe and each of its sub-sectors is an entire planet that takes a lifetime of experience to fully know.

To try and understand not just each sub-sector, but also synthesize between them and in conjunction to the economy and the environment is an even greater challenge.

The difficulty rises further since there is overspecialization in our times and there are few multidisciplinary experts to help us.

Regardless, it is more vital than ever to attempt it. Energy has long stopped being the domain of specialists and becomes more and more an issue for normal people, since we are talking about consumers who turn into producers through solar energy and batteries, vehicles that offer energy to the grid and participation in energy communities.

The crisis creates a more pressing need for everyone to engage with the energy sector, in order to make the right decisions for oneself and to support the policies of his/her choice on a national level.

2. What exactly is energy?

A) *The important basics*

It is weird, but in physics there is no commonly accepted definition for energy. Different scientists will give you different answers, but they all agree that energy is anything that produces work.

Humans use energy in two fundamental ways: Inside their bodies through food and much more outside of their bodies through exploiting all available sources: Sun, wind, water, coal, oil, natural gas, uranium etc, which we extract and use through our inventions.

Speaking about the energy sector itself, it is divided in two great categories: Fuels and electricity. A fuel is any liquid, solid or gas that can be put inside a tank, transferred through a pipe, a ship or a truck and finally get burned in a boiler or engine to produce heat or electricity.

Electricity on the other hand, is a continuous flow of electrons in cables and circuits that cannot be stored as easily as fuels. Electricity is more of an energy carrier and not an energy source, while fuels may play both parts. Fuels and electricity finally lead to the desirable work, whether it is motion, heat, cooling etc.

We often use energy sources, such as oil, gas, coal, the sun and wind to produce electricity, but the opposite happens only in specific cases, such as electrolysis for the production of hydrogen fuel from water.

The fundamental difference between fuels and electricity are a basic factor that defines what is possible and what is not possible in the energy sphere and the natural result of this balance is the so called "primary energy mix", which is the sum of all our energy sources. Globally, the primary mix consists of 31% oil, 24.4% natural gas, 26.9% coal, 4.3% nuclear energy, 6.8% hydro and 6.7% renewables[1].

Diagram 1: The primary energy mix

Shares of global primary energy
Percentage

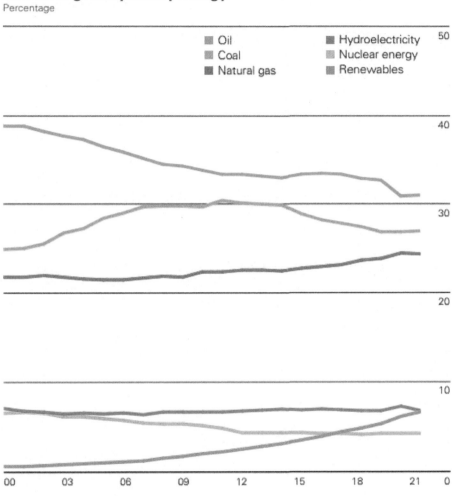

(2021, *BP Statistical Review 2022*)

Myths and truths: The primary energy mix

It is very important to remember the difference between the electricity mix and the primary mix, because there are many misconceptions and misunderstandings. All it takes is to keep in mind that the first is a subset of the second and amounts to about half the size.

We often hear for example, that "renewables reached 30% of production in country X", but this statement only concerns the electricity mix, not the entire energy of the country. Quite often, this misconception between the two mixes becomes the subject of exploitation by politicians and businesses who want to impress an audience that does not know the difference, therefore we need to be careful.

In general, each energy source has its own appropriate uses and often they overlay each other. For example, you can produce electricity with lignite, gas, uranium, renewables or diesel. You can warm your home with electricity, gas, geothermal energy etc. Some other times there are walls between energy technologies. For example, you cannot operate a large aircraft with solar energy.

What often defines where each energy source will be used is two things: Its facility of use and its energy density. In the first case, it is not rational economically and environmentally to transfer coal to each house in a city for heating, but it is easier to transfer natural gas or oil and even easier to use electricity. In the second case, a characteristic example is that oil has an energy density multiple times bigger than batteries. An aircraft, for example, needs that energy suddenly and at once and batteries cannot do that in the required scale, while they also have a large weight per unit of produced energy compared to liquid fuels.

If we examine the issue in historic terms, humanity always discovered new energy sources, however they did not replace one another, but they were added on top. When coal was discovered in the 19th century, we did not stop using biomass (burning wood). When we started to extract oil and gas, we did not stop using coal. When we created nuclear plants, we did not stop burning oil and gas.

The basic attribute of our energy history is that we constantly moved towards sources with a larger energy content and ease of use. Oil is much more efficient that coal and much easier to move and store. Uranium has a much higher energy density than oil, although its use is harder and demanding and at the end it was the balance between these two things that made it viable through the addition of human intelligence that found the best ways to utilize it.

The historical course towards "better" energy sources reached its end in the 21st century. Since climate change and environmental degradation raised awareness in – mostly – western societies, suddenly the goal became not just to cover our energy needs, but to do it in a sustainable way and reduce fossil fuel consumption.

As a result, for the first time we were forced to add new energy sources with clear disadvantages in terms of energy content and efficiency compared to older ones. Solar and wind energy produce electricity, but they do it in an inconsistent way, since they depend on weather conditions. Furthermore, they can mostly produce electricity so they cannot replace fuels at a large scale, at least not yet. And fuels are about half of the energy "pie", therefore they matter.

Even more important for our energy history, we are called for the first time to replace previous energy sources with these new ones. It is, of course, a gigantic challenge, since the entire modern technical civilization was based on and became possible as a result of fossil fuels.

If one examines economic growth in the decades after WW2, he/she will realize that it coincided with the era when oil extraction had the highest degree of economic efficiency. This is the so-called EROI, or how much energy you get from the ground for every unit of energy you invest. Moreover, during the decades from 1950-1990 the biggest discoveries of new deposits were made and after 2000 there was a drop with the exception of shale oil and gas, which had limited geographical application and specific prerequisites.

Another crucial element is that the high economic growth that took place after the war thanks to fossil fuels completely ignored the so called "externalities", meaning the cost of creation for fossil fuels (which is zero, since nature gave them to us) and the cost of their pollution. We simply had to spend what was needed for their extraction and transport and that is how we moved forward.

As we are going to realize later on in the book, the ignoring of these externalities plays a crucial part in our current problems.

How modern electricity production works

In order to produce electricity and cover demand, a series of production plants are used today to cover different needs. Firstly, we have base plants that are often hydro, coal and nuclear stations. Their defining feature is that they have a very predictable production so you know exactly what to expect. These plants cover a significant part of demand and renewables are added on top of them. However, renewables do not have a predictable production, but it ranges over time.

In order to plug the "hole" when renewables are low, we usually employ natural gas plants to balance them. The feature that makes gas plants ideal for this job is that they have a very short response time, so they can be turned on and provide energy within a few minutes, something crucial for the grid when renewable production falls. On the contrary, coal plants take many hours to get online, so they cannot play this particular role in the system.

Power production is shown in the following chart, where we can see a typical course over a few days. Solar production peaks in the middle of the day and it turns to zero at night, while wind production is greater in days with high winds and other sources are called to fill in when necessary.

B) Numbers that show our energy ignorance

The fact that we take energy so much for granted today is shown by how easy we are used to acquiring it. Electricity is the only good apart from water that anyone has access to simply through a hole in any wall of any building in the world. The average person is never further than a few feet away from a power supply point. Imagine how strange it would look if we could acquire any other good in this way.

Also, our ignorance is notable about how much energy we use in our lives. Almost no one knows for example, how much energy it takes to cultivate wheat. Furthermore, things that we have come to consider as immaterial actually have significant energy and climate cost. For example, the gadgets and networks we use for the internet reached 3.7% of global CO2 emissions in 2019, as much as air travel, while by 2025 they are expected to double[2].

As Nate Hagens, former VP of Lehman Brothers, who is now a professor of ecology economics, says, the energy content of fossil fuels is absolutely remarkable. A gallon of gasoline, when put inside a machine, produces in a few minutes the work that a human laborer would need a month to reach. This means that the 5 billion of human workers alive today have constantly at their side another 500 billion of imaginary workers who work day and night to produce real work for them. There was never a time in the past when the average person had this privilege, since it was enjoyed only by the nobles, lords and kings of the past.

If we could turn inside our minds all the fossil and non-fossil energy that we consume into barrels of equivalent oil and visualize it in real time, it would be like watching the Niagara Falls with oil instead of water.

As Jorir van der Schot says:

"Imagine you're standing at the side of the American falls at Niagara, like the people in the picture above. You can hear the roar, feel the thunder and see this enormous mass of water flow over the edge second after second after second.

Now imagine that instead of water flowing, it's all oil.
Such is the scale of man's societal metabolism: if you add up all of our primary energy supply and convert it to tonnes of oil-equivalent, like they do at the International Energy Agency, you'll find we're living on a *Niagara of oil.*"

Most of us were born inside the age of fossil fuels, therefore we have not experienced what came before in order to compare. Is it natural to consider all these rich energy sources as given, but it is also wrong. We consume fossil fuels like they are an inexhaustible source, while they are in fact non renewable and their margins are decaying constantly, especially as demand rises.

The Shah of Iran is said to have declared once that "oil is too precious to burn", referring to its myriad of other uses in the economy. This is another aspect that highlights our energy ignorance, that is how much energy and fossil fuels participate in processes we do not consider purely "energetic".

Vaclav Smil highlights that modern food production relies heavily on fossil fuels, through fertilizers made by ammonia and natural gas and used globally. It is their contribution that covered nutritional needs of 7 billion people in 2019, versus just 890 million in 1950. Ammonia, plastics, steel and concrete require around 20% of global energy and are responsible for 25% of CO2 emissions[3].

C) *The relationship between the economy and energy*

Since energy is after air, food and water, the most vital human need, it has a direct relationship to the economy.

In a certain sense energy is not a sector of the economy as we use to say, but the economy is a sector of energy, since it very strongly affects growth, prices, competitiveness and other critical indicators.

This becomes evident if we examine the relationship between the price index and energy. In the U.S. and Europe energy is about 9.1-9.5% of the price index, which means that the average citizen spends this amount of money to cover his/her energy needs. If prices in this share suddenly rise tenfold, then all the goods and services in the rest 90% has to drop by 20% in order to keep inflation steady.

This is what happened to us in 2021-2022. The wholesale cost of natural gas and electricity rose up to tenfold in many European countries, while oil and coal also rose significantly. Prices of the rest of the goods obviously did not fall, therefore inflation took off.

Energy is one of the basic pylons of the economy and affects its course to a large extent. Again, it is not an accident that the periods of the highest economic growth coincided with periods of cheap energy and discovery of new sources. It is also not accidental that the global economy is faced with a downturn during this energy crisis.

Hagens notes that in reality money itself is a claim for a quantity of energy, as it is included in every good and service in our economy. We spend one euro and we always receive something that took energy to create. Accordingly, every euro of debt is a claim for future energy production and consumption. "We can print new money and new debt, but we cannot print new energy", he says.

On a similar note, the Russian president, Vladimir Putin, said in September 2022 that "Westerners always print new money, but you cannot feed people or give them energy with paper."

Furthermore, energy and particularly electricity is a commodity with very low elasticity in the short term and so it takes a big rise in prices to balance the market, especially when there is a negative supply shock like the one we are witnessing today.

Another interesting element has to do with energy's relation to production that is energy intensity or how much energy it takes to produce a unit of

GDP. One would assume that as energy efficiency is improved and energy per unit of production falls, total energy consumption would fall as well. However, the exact opposite is true.

According to IEA, global energy intensity has fallen from 7.1 MJ per unit of GDP in 1990 to 4.7 MJ in 2018. During the same period energy consumption rose from 106,638 TWh to 172,884 TWh. The same trend is true for the last 200 years and there is a directly inverse relationship between these two values.

The inverse relationship between energy intensity and total consumption points to the Jevons Paradox, a phenomenon that appears insistently in the energy sector. In the 19th century, British inventor James Watt produced a new steam engine that needed less coal, therefore it had better efficiency. The goal was to reduce coal demand, which had reached high levels in the country.

Regardless, economist William Jevons found out in 1865 that in the years that followed Watt's invention, the exact opposite happened: Coal demand rose even further. The reason is that new and more efficient engines allowed the technology to spread to other fields of production than before, thus increasing coal consumption. Jevon's Paradox appears in many ways in the economy and another example is transportation: In the 1970s the U.S. adopted stricter fuel consumption requirements for cars in order to reduce demand, but drivers simply drove more miles with their new and more efficient vehicles.

It is something we see all the time in the economy. The promise to disconnect GDP from energy consumption and therefore from emissions is hampered by the fact that more and more products are made with a small life cycle. Each one may require less energy to manufacture and operate, but their very quantity cancels the energy saving effort.

D) The particularities of the energy transition

From 1980 onward, climate change started to alarm the scientific community. As recent reports showed, climate scientists were not the only ones that knew that fossil fuels cause global heating. Big oil companies also knew about the phenomenon since the beginning of the 1980s and they made sure to play it down in order to protect their activities[4].

At the beginning of the new century, climate change started to enter the forefront and became a topic in public discourse, resulting in the first serious policies about renewables growth, energy saving and reducing CO2 emissions. During previous decades there were many pilot programs for photovoltaics, wind energy and other technologies, but they could not compete with conventional technologies yet.

It would take a grand technological effort and primarily large capital and investments in order for "green" technologies to gradually mature and reach the level of directly competing against fossil fuels and then surpassing them. It took political will and pressure on behalf of citizens to make it happen.

Primarily after 2008, the EU and its member states made a great turn and decided to enforce massive new programs that would reduce CO2 emissions every decade in order to finally reach climate neutrality by 2050. Intermediate goals were set at the end of each decade in three different categories: CO2 emissions, energy saving and renewables penetration.

In order to succeed, they proceeded initially with handsome and often exaggerated subsidies (so called tariffs) for renewable technologies in order to grow them and make them cost effective in the long term versus fossil fuels through economies of scale. Indeed, the cost of photovoltaics and wind was dramatically reduced in the 2010s and the billions that European taxpayers paid led to a renewable cost already lower than fossil fuels before the crisis came. Essentially, Europeans paid the money, the Chinese made the factories (especially in solar) and the sector grew globally. Out of all the green technologies, the most successful was photovoltaics, since their cost was reduced by 80% from 2010 to 2019 (from 378$ to 68$ per MWh) and they became the cheapest and most competitive new energy source today[5].

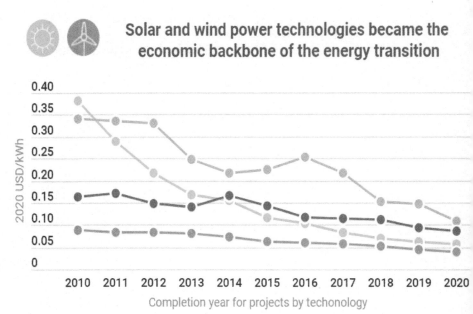

Solar and wind power technologies became the economic backbone of the energy transition

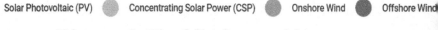

Diagram 2: The fall of renewables cost

(International Renewable Energy Agency - IRENA)

Another interesting aspect of the European effort was that essentially Europeans paid the price in order for all other countries in the world to have cheap renewables, even those who lagged in their climate ambitions. This is something that is not recognized enough today, when other countries scoff at the Europeans for their energy decisions and policies.

However, the energy transition had at its core the promise of disconnecting GDP growth from energy demand. This means that the economy will be able to continue to grow at the rate that we were used to (on average around 3% per year) while CO_2 emissions would fall. This is what is widely known as "sustainable development".

In order to assess the success of this commitment, only two things are important: What is the "pie", meaning the primary energy mix, and how it grows over time, meaning total energy demand.

In order for renewables to make a difference, they need to not only replace new added energy production, but also the existing one. Otherwise emissions remain stable at the best case and they are increased at the worst case. In the following two charts we can see the global electricity generation first as a percentage of each technology and then, in absolute numbers over time:

Diagram 3

World electricity generation by source
Percentage share

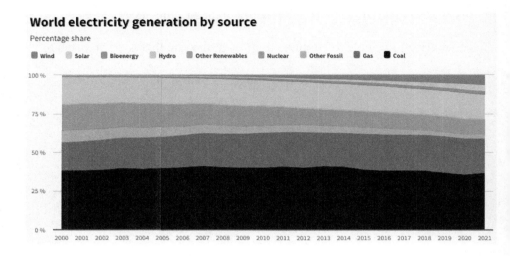

World electricity generation by source
Terawatt hours

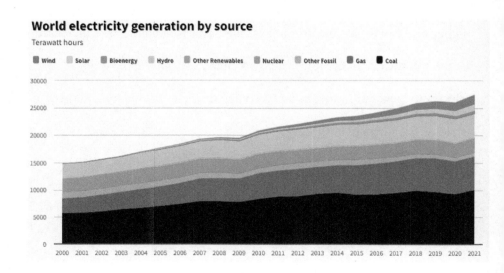

To put it in simple terms, if coal has a share of X in the energy pie and that share remains stable or falls slightly in time, while the pie itself grows rapidly, then in absolute terms consumption of coal (and its emissions) may remain stable. The same is true for oil and gas. Further down we are going to examine the European and global effort with this criterion in order to draw conclusions about the success or failure of the energy transition.

Another central idea is that renewables produce electricity more easily than they produce fuels, therefore it makes sense to grow the use of electricity in the primary energy mix against fossil fuels as much as possible. In this way, we will be able to make the mix "greener" faster and cheaper. This is called "electrification" and has its own set of challenges.

It should be noted that one country's success in the energy transition does not necessarily mean success in the wider climate effort if they are canceled by stagnation in other countries. For example, if a nation manages to reduce its emissions through transferring its factories in another country, then its emissions are simply exported and the first country appears to make illusory progress with no positive global results.

The same is true about energy sources. If Europe ceases to use coal, but it imports electricity that was produced by coal from third countries, then the planet makes no progress. This is why any good climate policy must have a global, or at least a regional character and the EU is trying in this regard with mixed results.

It is notable that Europe reduced its emissions from 3,2⁶6 million tonnes of CO_2 in 2000 to 2,651 in 2019, but China increased its own from 3,138 to 9,178 during the same period.

We should add another significant variable that is CO_2 emission rights and any carbon tax that came into force more widely during the previous decade as tools to discourage investment in "dirty" energy sources.

Europe was the leader here as well, since the EU was the first to enforce a wide emissions trading system, the ETS. The system requires from certain industrial sectors, such as power production, to pay money for every kilogram of CO_2 they emit during their operation. In this way the competitiveness of coal and partly gas (which is cleaner) is further reduced and renewables become even more competitive.

Shortly before the crisis, there was an overhaul of ETS that reduced available quantities of emission rights to polluters and pushed their prices up to around 80 euros/ton, a fact that played a role in the overall rise of power prices.

In terms of disconnecting economic growth from GDP, the EU has made the most progress globally so far. In its case, total energy consumption was reduced slightly in 2019 compared to 2000, while energy intensity was reduced significantly and economic growth was increased.

Regardless, European progress occurred largely as a result of turning its economy towards services and light industry, while the heavy industry was picked up by dirtier economies with higher energy intensity, such as China. Europe reduced its energy consumption from 2000 to 2019 from 1,471 million tonnes of equivalent oil (mtoe) to 1,403 mtoe, but China increased its own consumption from 1,144 to 3,403 mtoe. Again, we are witnessing the importance of the global aspect, as we did in the case of CO_2 emissions.

Internationally, fossil fuels cover around 83% of today's primary energy needs, versus 86% in the year 2000. Renewables cover less than 6% of primary energy. As Vaclav Smil notes[6]:

"What are the chances that after going from 86% to 83% during the first two decades of the 21st century the world will go from 83% to zero during the next two decades? Especially as a few weeks ago China announced additional 300 million tons of new coal production for 2022, and India additional 400 million tons by the end of 2023. We are still running into fossil fuels, not away from them".

The International Energy Agency (IEA) produced in September 2022 its own projection about where renewables must go as part of the primary mix by 2030 in order to maintain the course towards climate neutrality[7]:

<u>Diagram 4</u>

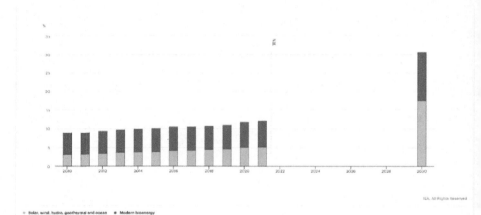

○ Solar, wind, hydro, geothermal and ocean   ● Modern bioenergy

(IEA, *Tracking Clean Energy Progress,* 2022)

Myths and truths: Subsidized renewables

People who are against renewables often say that they are energy sources that exist solely as a result of the subsidies they receive in various forms.

The reality is different: Renewables are now more competitive than fossil fuels even with no subsidies, as they can participate directly in the power market with no support mechanism. Since their cost was dramatically reduced in previous years, they have stopped requiring the high subsidies they once got.

One must also consider while comparing renewables to fossil fuels that oil, gas and coal are recipients of enormous subsidies, direct and indirect. The IMF calculates that global fossil fuel subsidies stood at 5.9 trillion dollars in 2020 that is 6.8% of global GDP.

3. How did we get to the energy crisis?

Journalists like to point out that the energy crisis is the result of a "perfect storm", where many negative factors coincided to bring things to an extreme point.

It is a fact that many unforeseen events and developments took place that affected the energy system and increased prices. But the crisis is also the result of more long term trends and policies that could have been partly anticipated.

There are four main reasons for the energy crisis: a) The market's planning, b) the chronic under-investment in hydrocarbons, c) the geopolitical fight between the West and Russia and d) the fall of nuclear energy.

Apart from the above, there are also plenty of unanticipated factors that affected things and aggravated the crisis, such as France's troubles with its nuclear plants and Europe's lack of interconnections that created choke points and limited ability to transfer natural gas from country to country.

At the same time, increased CO_2 cost contributed to the rise of power prices, while extreme weather, like intense drought in Europe and China during the summer of 2022, restricted power production in hydro and nuclear plants.

A) *The market design*

The beginning of the crisis has to be sought around 20 years in the past, when the architecture of the energy market in Europe was drawn.

Europe accepted, for better or worse, the ideals of the free and common market with the goal of having competition between producers and suppliers and the best possible prices for consumers.

Contrary to the past, prices were now formed not through regulation that is through a single dominant company or government decree, but from exchanges and the contracts of the many players in the new market.

But competition was not the only criterion, since under pressure from climate change, Europe wanted to decarbonize its energy as time passed. The goal of climate neutrality started to affect Brussels decision making from 2008 onward, leading to various different initiatives.

Thus, in Europe the term "marginal cost" was introduced since the beginning as the underlying value that set the wholesale price of electricity. This means that if each day there are for example, 100 renewables plants, 2 coal plants, 3 gas plants and 3 hydros in a country's system, each one has a different production cost because they have different costs of fuel, capital etc. This is reflected in the offers that producers make in the energy exchange to sell their production as they must cover their cost and also create some profit.

However, the wholesale price is set by the most expensive plant or, more specifically, the last accepted offer to cover demand for each hour of the day. If renewables offer electricity for 50 euros/MWh, lignite for 200 euros and gas for 300 euros, then all plants will be paid 300 euros and that is the price that consumers must pay.

It sounds irrational, but if you think about it, it makes some sense. The EU's goal was to promote renewables as much as possible and for investors to have the maximum incentive to build renewables and not conventional plants. What is better for an investor? To build a plan with an operational cost of 300 euros and get paid 300 euros or to build a plant that costs 50 euros and gets paid 300 euros? In this way the share of renewables would continue to increase compared to fossil fuels, thus achieving climate goals, which indeed occurred.

This architecture had its problems and certain mending was required, because when natural gas prices were low, the wholesale power price was not enough to provide the required investment signals in favor of renewables. Therefore, a support mechanism through tariffs was first introduced for wind and solar in order to be able to take off (when their cost was still high) and afterwards a system of auctions where projects with the lowest offers acquire capacity and are paid something on top of the wholesale price as a reward. There are also Power Purchase Agreements (PPAs) directly between producers and consumers with no intermediaries that gain traction in recent years. Another solution are Contracts for Difference (CfDs) where a certain price is set for renewables and when the wholesale price is higher or lower, the producer gets to return or claim the difference.

As analyst Jerome a Paris notes:

"The current market pricing system, based on spot prices, hinders investments in renewables, and investment has only been possible because workarounds have been put in place, like feed-in-tariffs in the past, and contracts for differences more recently. These regulatory regimes allow renewables, which have high fixed costs (linked to high upfront investments) but low marginal costs, find a price regime based on stable long term prices rather than volatile short term spot prices that can bring them to their knees in periods of low short term prices, even if their average long term price is competitive. These solutions are partly outside the "normal" market and are regularly criticized as providing "subsidies" to renewables even though they generate rather low long term prices (like <40 GBP/MWh for offshore wind in the UK), so it is absurd to say that "the current market provides the basis for investment"[8].

For many years during the past couple of decades, the price of natural gas was low and its use was encouraged. These years everything was normal and gas consumption increased in many countries around Europe, while plenty of gas plants were also constructed. Consumers of every category enjoyed low prices, while renewables were supported through tariffs and they also grew. Countries like Germany managed to have a competitive industry and low cost for its consumers.

However, the party was not meant to last for long. The market looked a bit like Chernobyl's nuclear reactor: It worked fine in normal conditions, but it would have great problems if conditions became extreme. There were no safeguards to protect against extreme cost rises, nor enough supervision for speculation and consumer protection. Even more important, under extreme conditions the market ceased to reflect the true cost of energy in a sense.

A basic element in the market's design was its dependence on the price of natural gas, as it is reflected in the Dutch contract TTF, which is the benchmark in Europe. However, TTF produces prices that do not take into account each country's particularities or different import prices. At the same time, it takes into account the various choke points between individual markets, since Europe is not a homogeneous system in natural gas, but there are areas with fewer or more interconnections and fewer or more sources of supply. This extra cost is embodied in the TTF price and even affects countries that may not face such issues. Lastly, TTF is a commodity with high liquidity and open participation, where thousands of players have access, therefore it can become the object of intense speculation.

During previous years, another important change also took place in Europe concerning TTF. As Russian gas was for many years more expensive than other sources, Gazprom's clients applied pressure to renegotiate their contracts, which were linked to the price of Brent oil until that moment. Now, clients wanted instead to connect their contracts to TTF, which was cheaper, and Gazprom indeed accepted their demands and new contracts were signed following the new logic. This fact would backfire on European consumers and their gas suppliers from 2021, when

TTF's price started to rise.

At that moment, the entire energy market was based on TTF and therefore, the rise in gas prices was passed on directly to electricity prices. TTF did not differentiate between countries with good access to low cost gas and countries with no interconnections and limited choices. It set one price for all and everyone's electricity was moving in the same way.

B) Investment in oil production

Another deep cause of the crisis has to do with fossil fuels and particularly hydrocarbons, meaning oil and gas. In the years after 2010, the turn from hydrocarbons to new energy sources was accelerated significantly, leading to very limited incentives for investment in new oil and gas production.

The global decarbonization effort led mainly after 2015 to a new investment culture, where more progressive investors and banks were unwilling to finance fossil fuel projects, such as oil exploration, gas pipelines and lignite plants.

Therefore, the oil and gas sector found itself faced with restricted capital during a period of low oil prices after the middle of 2014. In simple terms, the sector could not cover its needs through its own means or through bank financing and it had to curtail investments in order to survive.

Then came the pandemic to shake things up in the most dramatic way. The great reduction of economic activity in 2020 led oil prices even to negative territory (-40$ for WTI), a fact that shocked oil companies.

Immediately, they all entered a financial "straitjacket", reducing every cost in order to survive and avoid bankruptcy. They not only had to decrease investment in new production, but also curtail existing production, which the planet simply no longer needed.

As Bloomberg notes[9]:

"For the companies that survived the wave of bankruptcies, the business model had to change. Investors, burned by more than a decade of poor returns, wanted to finally see some cash."

Therefore, oil companies came up with business plans that had the goal of convincing any bank still willing to finance them, as well as creditors who wanted their money back. These included strict investment discipline, cutting jobs, selling assets and other moves that would allow them to survive, even though they worsened their future growth. In one word, austerity.

By 2021, the pandemic theoretically reached an end and economic activity sharply rebounded. The planet once again wanted oil, gas and coal, almost as much as it consumed before the pandemic.

However, oil companies could no longer increase their production to previous levels since a) they had pledged to the banks, b) it is hard and costly to turn oil wells on and off and c) they were afraid that the economic bounce could be short lived. Also, the cost of drilling started to rise as a result of higher metal and equipment prices. Additionally, the former lack of investment in exploration started to apply pressure to global oil production. In any case, oil companies had no economic reason to increase production, since at the new higher prices they had great profits even with reduced quantities.

It is notable that investments in new hydrocarbon production decreased from 500 bil. dollars in 2019 to 350 bil. in 2020 and a bit over 400 bil. in 2021[10]. The International Energy Forum believes that annual investment must reach at least 523 bil. in order to support current levels of production globally. In essence, this means that in oil we have been using what was discovered previously for a long time now and we are not replacing it with new discoveries.

The combination of these factors led oil prices to over 100$ for a large part of 2022 and to shortages that were mostly covered by OPEC+ production increases (the traditional OPEC plus Russia). However, OPEC+ had a hard time to reach necessary volumes and important members, such as Saudi Arabia, had to use part of their excess production capability that they always save for extreme circumstances. The fact that there was now little excess capacity to cover any new rise in demand was crucial for crude oil's price increase, since it filled the market with unease and uncertainty. Things were moving towards the edge and stock markets knew it well.

Furthermore, sanctions against Russia created a new landscape in the global oil trade with imbalances and rearrangements. Now, a larger part of Russian production is channeled to Asia and the West had to replace it with other sources of supply.

If we take a look at oil outside the strict bounds of this crisis, there are still plenty of deposits underground, but their extraction becomes more and more expensive in conjunction to the oil companies' available capital. However, oil remains crucial for the world economy and so far there is no alternative energy source with a wide application to replace it and cover every need.

In the case of oil, the goal is for supply to correspond to the – supposedly – reduced demand over time, something that has not occurred yet. On the contrary, the oil sector is faced with abrupt lurches and the lack of investment will become more evident from now on, unless something changes.

Myths and truths: Peak Oil

In the past, every time the price of oil increased greatly there was a lot of talk on the sidelines of the energy sector about the so-called Peak Oil, which is the moment when oil production will reach its maximum as a result of reserve depletion and the inability to find new ones. This discussion began during the oil crises of the 1970s and was repeated around the time of the 2008 crisis, when oil reached its historical high of 147 dollars/barrel.

In the years that followed, the discussion was moved from "geological" peak oil to the "economic" peak oil, which is the moment when its use will become either unwanted or uneconomical and will result in the historical wane of its consumption. When the pandemic appeared, demand fell dramatically to 91 mb/day and many experts and members of the energy sector, as well as journalists said that peak oil had already occurred or is expected to come in the next few years.

Based on what happened so far, these projections about demand peaking in 2019, when it reached 99.7 mb/day are premature. IEA foresees that demand will reach these levels again at the end of 2022 and will exceed them in 2023 with 101.8 mb/day, while in 2026 it will reach 104.1 mb/day

The result of the energy crisis on oil remains to be seen in order to draw conclusions. So far, the prophecy of former CEO of Total, Christoph de Margerie, stands true. He said that global oil production will never exceed 100 mb/day.

C) EU and Russia – The geopolitical catalyst

In natural gas there are some common themes with oil, since these two hydrocarbons are often produced from the same deposits. The lack of investment is an important factor in natural gas that led to higher prices. However, it is their differences that are more pronounced when it comes to this energy crisis. It is not so much a lack of natural gas that led us here, but far more the exclusion of a large part of it geographically.

The basic cause for the rise of natural gas prices was therefore geopolitical and has to do with the war in Ukraine and the spat between the West and Russia, which had started much earlier. In effect, the war in Ukraine and political decisions made by all sides removed almost overnight a large part of the European system's supply, which cannot be easily or quickly replaced with other sources.

Russia used to cover around 40% of Europe's natural gas needs in previous years, mainly through pipelines and partly through LNG. In the middle of 2022 this percentage was reduced to 9% with a very clear possibility of reaching zero in the next months or years[11].

Diagram 5

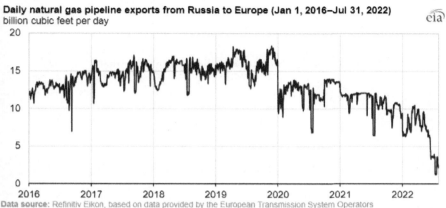

Daily natural gas pipeline exports from Russia to Europe (Jan 1, 2016–Jul 31, 2022)
billion cubic feet per day

Data source: Refinitiv Eikon, based on data provided by the European Transmission System Operators
Note: Russia's natural gas exports by pipeline include exports to the European Union and the United Kingdom as measured by daily flow volumes at the main entry points in Germany, Slovakia, and Poland.

(U.S. Energy Information Administration - EIA)

In order to replace the lost Russian energy, the EU draw ambitious plans to find gas from other countries, to advance cheap and clean renewables, to save energy and other measures. The 210 billion euro REPower EU plan, which was presented in May 2022, calls for renewables penetration of 45% by 2030 from 40% earlier and it includes a European Solar Strategy for the doubling of PV by 2025 and the installation of 600 GW by 2030 with mass new installations on rooftops and public and commercial buildings. At the same time, it calls for doubling the rate of installing thermal pumps, to accelerate licensing for renewables, to produce 10 million tonnes of green hydrogen and import another 10 million tonnes by 2030, as well as producing 35 bcm of bio-methane by 2030.

As a result, the replacement of fossil fuels is expected to save up to 35 bcm of natural gas by 2030, far more than the previous plan "Fit for 55" projected.

However, these efforts take time to make a difference and they are not enough to cover the huge gap that lost Russian gas created. This is why the EU also proceeded with emergency energy saving measures (-15% in power and gas consumption for the winter of 2022-23), as well as maximizing imports from other regions to plug the hole.

D) *The lost nuclear energy*

Fukushima's legacy

The nuclear accident that took place in Fukishima's power plant in 2011 was the second largest emission of radioactivity in history after Chernobyl. In contrast to Chernobyl, the planet was much more interconnected in 2011, while the accident did not take place in a secretive country, such as the Soviet Union, but inside an open and liberal nation.

This meant that the flow of information and images from the accident made their way everywhere and they caused great impression and finally, reactions. Gradually inside the next few months and years, public opinion in countries with high nuclear energy, such as Japan and Germany, changed drastically. Now, most citizens were against nuclear power plants operating on their own soil and pressured their governments to stop investing in nuclear plants and even withdraw existing plants.

The result of this turn was that the German government pledged to decommission all nuclear reactors by the end of 2022 based on a gradual schedule. As part of the German energy transition policy, called

Energiewende, nuclear plants would be decommissioned first and coal plants would follow at a later stage, while renewables and natural gas plants would be called to balance the lost output step by step.

The German chancellor at the time, Angela Merkel, said in 2011:

"We can become the first industrialized country to abandon nuclear energy. It is a Herculean effort. If we turn faster away from nuclear towards renewables, then we are going to need fossil fuel plants for the transition."

Merkel's decision also had political incentives, since coal creates many more jobs in the country compared to nuclear energy, therefore it made sense to keep these plants running for longer.

The purpose of this book is not to judge whether nuclear energy is acceptable or not. This is a matter for the educated reader who is going to make his/her own conclusions. Regardless, nuclear energy has certain great benefits when it comes to climate and energy policy: It has zero CO_2 emissions, it is very reliable as base capacity and in the case of older reactors whose initial construction cost has been covered long ago, they have very low power production costs.

The gradual decommissioning of German nuclear plants was accompanied by an anti-nuclear culture in many other European nations, a fact that blocked the construction of new nuclear plants that would otherwise have been realized.

The result for Germany itself is that while nuclear energy had a share of 22.2% in its electricity mix in 2010, it reached just 11% in 2020 and before the end of 2022 only three plants with 4 GW were in operation.

France's role

Something unforeseen came to be added on top of all these problems. France, a nation that traditionally produces electricity primarily (80%) through nuclear stations for a very low cost, started to face crucial trouble with the safety of its reactors that had to be resolved at once. The French nuclear reactors had erosion in their pipes and tens of them were put offline in 2021 and 2022 in order to be repaired.

The result was that France temporarily lost around half its reactors and instead of a power exporting country that it was traditionally, it became strictly an importer. It is notable that during the summer of 2022 France often had three times as much imports as exports. This fact by itself led to a domino of power price increases across Europe.

The logic that explains this domino is simple: When the French price is higher than Italy's, then the first is going to import power from the second. Since Italian power will now be in higher demand, its price will rise, pushing up its own neighbors' prices and so on.

Indeed, this is not the first time that France has caused a price domino. Back in the winter of 2016-17 something similar occurred, but at that time the other factors of the crisis were not present so the price rises never reached dramatic proportions.

4. The "perfect storm"

A) *The chronicle of the crisis*

The pandemic years

The Covid-19 pandemic was a major shock for global supply chains, which were designed to operate on a razor's edge in order to maximize profits, with no significant margin for flexibility.

The sudden economic deceleration and demand-consumption destruction shook supply chains in every good and resource, as well as industrial production for a long time. During 2020 and 2021, many production lines, mines and factories had to shut down resulting in drastically lower demand for energy.

Moreover, investment plans of many billions of dollars were affected and they had to be reconsidered or canceled. Correspondingly, investment plans in energy, such as PV plants, had to suffer multi-month delays since companies could not acquire the equipment they needed in time.

Winter 2021-2022

The winter of 2021 was especially cold for regions such as Europe and Asia, a fact that raised demand for natural gas and its consumption. However, since supply was no longer in line with demand, the result was that levels of European gas storage, which is typically used during the winter, were steeply reduced.

Each year consists of two seasons for the needs of natural gas: The winter semester, when storage gradually empties up till March and afterwards the summer semester, when it is replenished bit by bit. During a typical winter, storage covers up to a third of natural gas demand and the rest is supplied by imports.

Since European storage fell in March 2021 at the level of 30%, lower than the last five year average, its replenishment automatically became harder during the summer season. Liquefied natural gas (LNG) was not enough to cover demand and Europe was faced with competition from Asia, who also had its own needs. Meanwhile Russia began from August to constrict the quantities of gas it offered through auctions.

We should note at this point that Gazprom provided gas to Europeans in two ways: Large quantities through long term contracts signed with its clients, but also smaller quantities in the physical market through auctions taking place every few months.

So the Russian tactic of gradually reducing gas exports started through auctions and during the fall of 2021 the situation in the market led the price of TTF to the level of 48 euros/MWh.

This was the first wake up call for the EU and the winter of 2021-2022 that followed showed that things were harder than what governments and Brussels had anticipated. Competition for LNG was intense, the Russians continued to cut their supplies in the auctions and European storage was being rapidly depleted. TTF reached 129 euros in December 22 and then was reduced to 70-80 euros, still a very high price compared to the past.

At the end, Europe was saved that winter primarily thanks to the weather, which was mild and allowed storage to finish the season at similar levels to the previous year. So it was that the worst scenarios were avoided.

The fateful 2022

However, in February 2022 the unimaginable happened and Russia invaded Ukraine. This fact changed everything, not just in energy, but in international politics.

The war multiplied a process that had begun years ago in the global stage and leads to a reversal of globalization, fewer trade relations than before and more isolation. The phenomenon was already happening through American sanctions versus China, Russia and Iran, but reached entirely new and unforeseen dimensions after the invasion in Ukraine.

Essentially, Russia suddenly stopped being a trading partner for Western countries, a fact that removed a huge quantity of energy supplies from Europe's available solutions. In other words, there was a Europe that was designed to function based on certain parameters and through political decisions, one of the most important stopped applying.

In a certain sense, this this is the whole point of the energy crisis: That it is not caused exactly by a lack of energy sources, but especially in natural gas by an exclusion of sources through political decisions.

The consecutive sanction packages of the EU against Russia were followed by respective reductions of Russian natural gas supplies, this time not through auctions – who were already at zero – but in the contracts themselves.

In March 2022, president Putin announced that payments for gas contracts had to be made under a new system that converted euros to rubles. "If payments are not made like this, we will consider it a breach of contract on behalf of the buyers and they will have to suffer the consequences. No one sells to us for free and we should not engage in charity, meaning that existing contracts will be halted."

The EU had a jittery reaction and at last it largely accepted the new system, while the handling of the issue by Brussels was problematic since it allowed it to become a cause of division and friction inside it.

When Putin's deadline arrived in May, Moscow completely cut supplies to Bulgaria and Poland who refused to pay the way he wanted. During the following months, more cuts would follow to other European states, such as Finland, the Netherlands and Denmark.

In an effort to respond, the EU Commission announced in May a big and ambitious plan in order to face the crisis, to reduce dependence on Russia and achieve its climate goals, the so called REPower EU.

This time, the EU would be ready for even the worst outcomes and would fight to maintain enough LNG supplies to make it through the winter. At the same time, it prepared for a gradual, but short term independence from Russian gas through growing renewables even faster, betting on energy efficiency and finding natural gas from other sources on a more stable basis. Also, at the end of May the EU announced a gradual oil embargo on Russian oil.

As the president of the EU Commission, Ursula von der Leyen said then:

"We must reduce our dependence from Russian gas, oil and coal. We cannot rely on a supplier that openly threatens us. We must act immediately to face the effects of higher energy prices, differentiate our gas supply for the next winter and accelerate the clean energy transition. The faster we turn to renewables and hydrogen, together with energy efficiency, the sooner we will become independent and masters of our own energy system."

Despite the emphasis on green energy, the EU realized that it is incapable of covering the gap immediately. During the summer months a fundamental change took place for the European energy trilemma that consists of "security of supply", "sustainability" and "cost". Before, the

EU focused more on sustainability, but now it would turn its attention to security. The political cost was huge and left no other choice to governments and the Commission. It was decided to cover energy needs at any cost. Europe would pay as much as it had to in order for its citizens to have enough gas and electricity.

With Brussels' blessing, the disgraced coal would once again be counted on to reduce gas consumption in order to save enough for winter. Additionally, diesel would replace gas for some time in power plants with a fuel switching ability if Russia completely halted its supplies.

In August, France's problem with its nuclear plants reached its maximum and Gazprom was turning the Nord Stream pipeline on and off, a fact that rose power prices to levels around 700 euro/MWh in countries like France, Italy and Greece. The TTF contract reached almost 350 euros/MWh.

Meanwhile, during the summer the capacity of the Nord Stream 1 pipeline was gradually reduced from 55 bcm in June to 36.5, then 24.5 and finally 12 bcm before it completely stopped in August. At the beginning of September Russia said that Nord Stream 1 would not get back online as long as sanctions were in place and Europe received only limited amounts of Russian gas through Ukraine and through Turk Stream pipeline in the South, as well as some Russian LNG cargoes.

Looking at the near future, it is entirely possible that even these volumes will cease, especially if Europe decided a price cap on Russian gas imports or if there is further escalation on a political level.

At the end of September, both Nord Stream pipelines were sabotaged and will remain inoperative without time consuming repairs.

B) Extreme prices in Europe

The energy crisis is often called a Lernaean Hydra, since it has many "heads" and it concerns concurrently the high cost of oil, fuels, natural gas, coal and electricity. Indeed, this symbolism becomes even more evident, since it is often that one commodity's price is reduced while another is increased during all this turmoil.

Let's examine first the price of the natural gas contract, TTF that Europe widely uses. At the beginning of September 2021 it was near 30 euros/MWh. Exactly a year later, it stood at around 230 euros, after having climbed all the way to 346 euros that is ten times higher in just one year. The rise and volatility in TTF is such that in a single day it may move up and down by up to 60 euros, twice its own price before the crisis. It is also impressive that the maximum price reached in TTF translates to over 500 dollars per oil-equivalent barrel[12].

Diagram 6: TTF

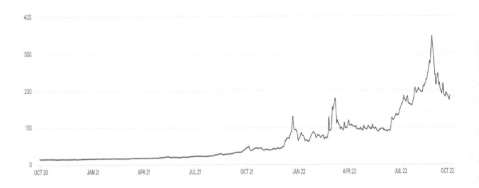

(Intercontinental Exchange)

As we noted, the effect of high gas prices to electricity prices was immediate. According to Eurelectric, wholesale power prices increased by 532% in Europe from January 2021 to August 2022. If we go further back to compare, in 2020 prices were on average at 34 euros/MWh and in August 2022 they reached 405 euros, therefore they, too, became ten times higher.

Retail prices on the other hand, were increased by 84% from the beginning of 2021 to spring 2022 in Europe.

As we noted, the effect of high gas prices to electricity prices was immediate. According to Eurelectric, wholesale power prices increased by 532% in Europe from January 2021 to August 2022. If we go further back to compare, in 2020 prices were on average at 34 euros/MWh and in August 2022 they reached 405 euros, therefore they, too, became ten times higher.

Retail prices on the other hand, were increased by 84% from the beginning of 2021 to spring 2022 in Europe.

Myths and truths: ETS is responsible for high electricity prices

The European emission trading system (ETS) was overhauled before the energy crisis and a significant amount of emission rights was removed from the market and placed inside a special reserve, thus limiting their availability to polluting industries. As a result, their price increases from 33 euros/tonne at the beginning of 2021 to 70 euros in September 2022, after peaking first at 97 euros.

Despite that fact, according to Eurelectric, natural gas is the culprit for the larger percentage of the rise in wholesale power prices and CO_2 emission rights to a much smaller degree. According to the association's data, in April 2022 the wholesale price was 214 euros/MWH, of which 184 euros were related to the cost of natural gas and less than 30 euros to the CO_2 cost.

Significant changes also took place in Europe's power production mix from the first half of 2021 to the first half of 2022, as natural gas was reduced by 6 TWh, nuclear energy by 33 TWh and hydro by 37 TWh. On the other hand, wind energy increased its production by 24 TWh, coal by 18 TWh, solar by 17 TWh and other renewables by 10 TWh[13].

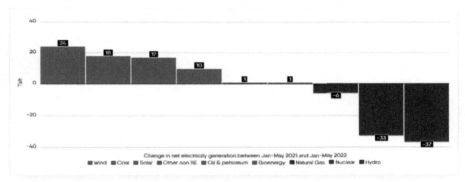

Change in net electricity generation between Jan-May 2021 and Jan-May 2022
Wind Coal Solar Other non RE Oil & petroleum Bioenergy Natural Gas Nuclear Hydro

Diagram 7

(Eurelectric, Power Barometer, September 2022)

As we can see in the chart, two very reliable and cheap sources for the system, hydro and nuclear, had an intense fall and this fact affected adequacy negatively while pushing up prices.

In countries like France, Italy and Greece, as well as the Western Balkans, prices reached even 700 euros/MWh in the wholesale market during August 2022.

C) *Winners and losers of the energy crisis*

As the energy crisis that hits the planet reached a crescendo in 2022, the first conclusions are available about who gained and who lost during this whole time.

Naturally, this estimation is based on the rise of oil and gas prices, the geopolitical battle with Russia, the pandemic and its effects on supply chains, as well as special circumstances such as drought and extreme weather effects.

Winners

Energy saving

During previous energy crises, such as the oil crisis of 1973 and 1979, energy saving played a large part, as even large economies like the U.S. and Britain applied special measures and guided their citizens to make voluntary moves to help.

These years there were speed limits in highways, fuel rationing, but also more long term decisions, such as stricter fuel consumption requirements for new car engines.

During this present crisis, energy saving measures have started being applied, albeit with some delay. They were presented largely after the middle of 2022, when it became obvious that things were moving to extreme directions.

Germany was the country with the greatest effort in this regard in order to save natural gas to be used during the hard winter months. Berlin's measures include among other things, that shops' doors need to be closed during the day, lighted ads will turn off after 10 at night, as well as lighting in monuments, while corridors in public buildings will not be heated and the thermostat will be set at 19 degrees in offices. Furthermore, the government enforces long term energy saving measures, such as inspecting every heating system in the country. All of the above are expected to save 2-2.5% of the country's gas consumption.

In Italy, new rules concern public buildings where cooling temperature in the summer may not be lower than 25 degrees and in winter not higher than 21 degrees. The operation of heating in public services, private offices and houses must be reduced by one hour daily. Also, it was decided to keep radiators on for 15 fewer days than last year. Similar decisions are being taken across Europe.

On a European level, the EU tries to forward energy saving any way it can, through voluntary action on behalf of consumers, or targeted interventions. The EU Commission also presented a new system that connects consumer subsidies to the extent of energy saving they achieve. Furthermore, there is already in place a mechanism of remunerating industries that choose or have to shut down during the winter months. All these are unprecedented and very telling of the size of the problems Europe is faced with.

Of course, it is unfortunate that states and citizens remember the importance of energy saving only during crises, since its support during "easy" times would greatly reduce the danger of having a crisis in the first place.

Whatever the case, governments around the globe are now encouraging their citizens to reduce consumption and set in motion ambitious energy saving plans for buildings, appliances and transportation with significant grants.

The problem is that these plans are accompanied by large subsidies in consumption that counterbalance demand reduction and simply postpone the issue.

Oil companies

Both private and state oil companies made record profits in 2022 thanks to the price of oil, which peaked at 123 dollars/barrel in March and spent almost the entire first half of the year over 100 dollars.

ExxonMobil had 17.9 billion dollars in profits during the second trimester, while Chevron had 11.6 billion, Shell 11.5 billion and TotalEnergies 9.8 billion.

According to Financial Times[14], the five biggest western oil companies made more than 50 billion dollars just in the second trimester of 2022, and they are added to their already high profits in the rest of the year and 2021.

Things are the same if we are talking about state companies, like Saudi Aramco, who had 48.4 billion in profits during Q2 and 87.8 billion during Q1 2022. Brazilian Petrobras made 9.1 billion in Q2, up from 7.7 billion one year earlier. Malaysian Petronas more than doubled its profits within a year, reaching 10.3 billion dollars.

There were also great opportunities for profits in various sub-sectors, such as diesel, which had its own rallies in the markets as a result of scarcity, especially in the U.S. where refineries were operating at maximum capacity to cover demand at the midst of an economic rebound.

Essentially oil companies did not have to do anything new during the past year, since they made far more money while maintaining the same production. Especially private and listed companies just watched their profits take off without investing in new drilling despite the calls by Western governments. However, the future may be different, since perpetual high prices will increase security for oil companies and may convince them to invest in production again.

LNG producers

American companies, such as Tellurian and Cheniere Energy investing significant capital during the previous decade in order to build natural gas liquification plants in the eastern shore of the U.S. with the goal of exporting. At the same time, their decision to embrace a more flexible commercial policy with a combination of long term contracts and spot cargoes proved prescient. The price difference between American and European gas was so high that it allowed them to make enormous profits and to schedule new investments that are expected to come online from 2023 onward.

Specifically, American Cheniere Energy saw its profits double to 741 million dollars in H1 2022, from having 329 million dollars in damages a year earlier. The other big producer, Tellurian, had 61.3 million dollars in profit against just 5.6 million a year before.

On a similar note, state producers like Qatar saw their value rise in the global energy scene and they are taking advantage of the geopolitical crisis. State owned Qatargas, one of the largest LNG exporters in the world, exported more than 16 bcf/day in April 2022, a five year high. This by itself increased Qatar's public surplus by 12 times to 12.8 billion dollars in H1 2022.

Traders

The extreme difference between oil prices in 2020 (when they reached negative territory) and 2022 (100$+) was just one example of how volatility turns to large profits for traders. Similar situations occurred in natural gas, electricity and various other commodities, where the losses of some players translated to a party for others.

Commodity houses, such as Glencore (who made 8.9 billion in H1 2022), have made record profits during recent years. It is notable that Gunvor, who specializes in the LNG trade, announced higher profits for H1 2022 than the entire 2021 (841 million dollars), which was already a very lucrative year for the company.

These commodity houses are not the only traders in the market. Oil companies, like Total, and utilities like E.ON and EDF have their own trading offices, while there is also a plethora of smaller players in this sector who had great opportunities to get rich by betting on the ups and downs of commodity prices.

Renewable energy sources

The cost of renewables has risen during the last year as a result of supply chain problems, inflation and the cost of materials. Regardless, it remains far lower than other power production technologies, something that this crisis underlines.

IEA believes that renewables have the ability to greatly reduce prices and dependence on fossil fuels, both in the short and long term. Although the cost for new photovoltaics and wind farms has risen and paused a decade of continuous reductions, the prices of oil, natural gas and coal increased much faster and this leads to green electricity being more competitive[15].

Renewables already have a great boost in Europe through the REPower EU initiative and despite the temporary revival of coal, they are expected to accelerate globally, as states realize that they are the only way to move forward and enhance their energy security. Apart from their cost, renewables offer geopolitical advantages as well, since the fuel is just the local sun, air and water, a significant fact for a Europe that does not want to be dependent on Moscow anymore.

Spain-Portugal

Spain and Portugal acquired at the beginning of 2022 the so-called "Iberian exception", which means that the EU Commission gave them the ability to disconnect the price of gas from the wholesale power price with positive results for their final consumers.

At the same time, that particular region has a large regasification capacity with LNG terminals and zero dependence on Russian gas, a fact that together with its limited interconnections to the rest of Europe makes it an "island" in essence and it protects it to a certain degree from the crisis in the rest of the continent.

However, the Iberian Peninsula's protection from higher prices may reach an end soon, if there is a unified solution on a European level that will remove exceptions.

Myths and truths: The Iberian model

In the spring of 2022, Spain and Portugal asked the EU Commission to be exempted when it comes to the way their own energy market works. Brussels agreed and so the "Iberian exception" was realized. From that moment onward, the wholesale power market operated based on a technically lowered price of natural gas compared to TTF and correspondingly, a much lower power price was formed. Consumers of the two countries had much better prices compared to the rest of Europe.

The issue with the Iberian model is that it could not be enforced if Spain and Portugal were not isolated from the rest of the continent. This occurs because they have limited interconnection to neighboring countries and thus, limited imports and exports of electricity.

If these two countries had extensive interconnections and realized this model, then the difference between their own price and the one in France would automatically lead to very increased power exports and an alignment between them.

Accordingly, France would partly gain by lowering its own price through cheap Spanish electricity, but eventually this, too, would realign through French exports to other countries, like Italy. The final result would be limited for everyone involved.

The only way to realize the Iberian model in other European countries would be to somehow enforce it in the whole of Europe at once. In fact, the EU Commission and member states were discussing this very plan in October 2022.

5) The repercussions of the energy crisis

A) The economic effects

The crisis so far had pronounced effects on the European economy, for households, small businesses and industries.

According to JP Morgan, absent policy intervention, households could see energy bills nearing 30% of disposable income based on spot prices, three times above the typical definition of "fuel poverty"[21].

Diagram 8

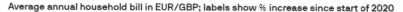

Average annual household bill in EUR/GBP; labels show % increase since start of 2020

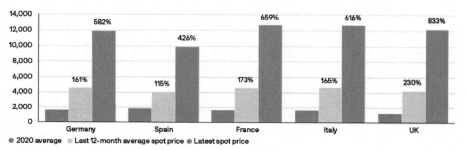

Source: Jefferies, J.P. Morgan Asset Management. For illustrative purposes only. Data as of 31 August 2022.

As European gas prices soared up to eight times their 10-year average, countries introduced policies to curb the impact of rising prices on households and businesses. These include everything from the cost of living subsidies to wholesale price regulation. Overall, funding for such initiatives has reached $276 billion as of August 2022[22].

High energy cost drastically reduces Europeans' disposable income. In the U.K. even with the 150 billion pounds price cap enforced by the government, the impact of higher energy costs still means real household

87

disposable incomes will probably now fall by roughly 2.5% in 2022 and by about 1.0% next year. The situation is far worse in poorer countries[23].

Households have the greatest protection against price rises compared to other consumers. In the case of businesses, the situation is worse. In places like Greece, small businesses saw power bills up to four or five times higher during H1 2022 and even after new subsidies they still have to pay a lot more than they did in 2021.

According to EuroCommerce, European retail businesses operate with very low margins of typically 1-3%. With companies facing a quadrupling of their energy bills, the cost of energy now accounts for over 40% of EBIDTA."As an essential service to customers and the rest of the supply chain, our sector cannot simply halt its operations to take account of energy prices", says EuroCommerce in a policy paper[24].

If things are hard for households and businesses, the effects are extreme in the case of European industry, forcing factories to cut production and put tens of thousands of employees on furlough, as industrial production in the euro area fell 2.3 percent in July from a year earlier, the biggest drop in more than two years[25].

According to the Eurometaux association, 50% of the EU's aluminium and zinc capacity has already been forced offline due to the power crisis, as well as significant curtailments in silicon and ferroalloys production and further impacts felt across copper and nickel sectors. Producers face electricity and gas costs over ten times higher than last year, far exceeding the sales price for their products[26].

Fertilizer producers were especially hit by high prices, since about 70% of the cost of producing fertilizer is the price of natural gas. Overall, fertilizer production in the EU has been reduced by about two thirds.

B) A leap for renewables, but obstacles remain

Investment interest for new renewables projects is rising as a result of the crisis in Europe, despite their rising cost.

The extreme rise of prices in gas, coal and oil make renewables even more competitive than before, however there are still obstacles to their development that need to be addressed.

The main issues faced by renewables in Europe are three: Grid constraints, public reactions and licensing.

When it comes to the grid, there is already congestion in many parts of European countries so they are not able to accommodate more renewables. The grid operators have planned significant new investments to upgrade, modernize and digitize their grid in order to increase new connections both in the transmission and the distribution network. The bet is for the grid to develop in step with renewable goals and not fall behind. This is not so hard in terms of financing, which is readily available, but more so in technical and organizational terms.

Another fundamental issue for renewables is the public often reacting to new wind and solar farms. It is quite often that we see local communities and organizations turning against new projects in their area and delaying their construction for years.

Renewables companies claim that they require some kind of assurance on behalf of the state in order to know beforehand whether they can complete a project or not. If they acquire the necessary licenses, they must be able to complete their investment without any obstacles, they highlight.

On the other hand, citizens and local communities often believe that the natural environment is degraded as a result of renewable projects, while other economic activities such as tourism and agriculture are threatened.

This is obviously a difficult conversation, but regardless one that will affect our energy future. The goal is to have some kind of balance between different pairs of public interest: The natural environment on one hand and facing climate change on the other. Also, promoting renewables or economic activities of other shorts where the creation of new jobs and local value is important.

The third major issue for renewables is the licensing process, since investment time until a project begins commercial operation is high and in the case of wind farms it may reach up to 7 or more years.

REPower EU aims to reduce investment time through a simplification of procedures, but it will depend largely on individual member-states since many of the licenses have to be issued by local bodies.

Myths and truths about renewables

Since we live in an age of intense conspiracy theories and fake news, many of them apply to the energy sector, especially renewables. It is hard to know the exact motives of those who are against their development because sometimes they are sincere and related to environmental and economic concerns, while other times they are the result of prejudice and disinformation, so they are dogmatic in nature.

In this last case, there are many people who simply do not believe in climate change and no discussion can be made here since there is a fundamental ideological gap. There are others who are against renewables for the wrong reasons, for example, they claim that there are threats to human health that simply are untrue.

The only certainty is that in a democratic society all points of view must initially be heard and no government can enforce projects without any terms or public dialogue. What is needed is a more essential discourse between the sector, public bodies and citizens, as well as adequate incentives for local communities who host renewable plants.

We should also note that European law sets renewable development as an issue of European and national interest, which overrides local interests. This will become a significant issue from now on and governments must treat it carefully.

Apart from the above, what should worry us more than anything is the dogmatic refusal of any possible solution: Many people do not accept renewables because they degrade the local environment, they do not want gas because it is expensive and imported, they refuse coal because it is dirty, but they also want cheap electricity. Obviously, there can be no such solution and we have to make difficult choices for the best possible result.

6. Three scenarios for what is to come

The course of the crisis and the end of relations between the West and Russia brings us face to face with an entirely new reality globally. Politics that were considered certain previously are now disputed, business plans must be revised, while lobbies are fighting hard in the EU to shape the future.

Already there are battles taking place in Brussels among different interest groups. On one hand there are fossil fuel companies that reminded everyone of their continued significance and on the other hand renewables that are portrayed as the only viable long term solution in terms of cost. There is also competition inside each sector, since even small changes in policy translate to big fluctuations in profits and play a large part in the viability of each technology and each corporation.

The EU Commission is now more open to different energy choices and this widens the array of possibilities and creates opportunities for acquiring relative advantages. Moreover, there is global competition with recalibrations and new dependencies between states.

On a consumer level, although everyone wishes for energy cost to return where it was before the crisis, most experts believe this is not possible in the short term. In contrast, we have entered a process where all consumers, small and large, will be much more active than before and will be seeking the best energy sources and the best products, as well as changing suppliers more often than in the past.

Obviously, no one can make exact predictions about where the crisis leads us. The variables are too many and uncertainty too large, so there are almost no sure bets. Therefore, we can only examine three distinct scenarios and in the end we may see some combination between them.

A) *Return to normality*

How would a path of return to the normality we used to have look like?

Of course, a major issue is the relationship between the EU and Russia, which is now close to a breaking point after repeated strikes from one to the other. Since the war in Ukraine has taken a more slow form, it is doubtful that there will be a winner in the near future. A more possible development that could lead to negotiations between the EU, Russia and Ukraine would be if Russia can no longer stand the weight of economic sanctions and decides to back down diplomatically, but this is hard because so far the Russian economy appears to be durable under pressure.

Likewise, it would be very hard for the EU to cave in to Russian demands after all the commitments it has made. Therefore, only regime change in Moscow would lead to a reset of Russian-European relations so it makes no sense to consider it as a short term scenario, although it is entirely a possibility in the long term.

So what else can happen to improve things for Europe? By the end of 2023 significant new LNG capacity will be available globally, with most of it in the U.S. where three new LNG plants are coming online and hopefully will improve supplies and lower prices for Europeans.

It should be noted that global LNG investment is expected to rise from just 2 billion dollars in 2020 to 42 billion in 2024, according to Rystad Energy. Many new projects are expected to come online in 2024 and 2025 and improve the situation, while compensating for lost Russian gas[27].

These investments include projects such as Golden Pass in Texas that will become operational in 2024 with 18 million tonnes annually, Plaquemines in Louisiana with 24 mta from 2025, but also Qatar, which aims to increase its own production gradually from 77 to 126 mta by 2027. Already there are discussions between European companies and LNG producers for signing long term contracts at a cost of just one seventh compared to current TTF prices.

Apart from LNG, the expected return of most French nuclear reactors in action during the winter of 2022-2023 is going to be permanent – barring any unforeseen delays- and will improve the country's capability to cover its own demand and help its neighbors. This was already shown by the willingness of president Macron and chancellor Sholts to help each other during the hardest months of the crisis.

Another promising element is the European effort through REPower EU and energy saving measures already in effect. As time passes, the results of these moves will be felt and they will gradually bring benefits, as today's pain may become a benefit tomorrow.

Since the crisis concerns oil as well, here too there is a possible positive development concerning Iran and the return of its exports in the international market. The U.S., Europe and Tehran spent all of 2022 engaging in intense negotiations about repeating the former nuclear deal. If there is a new agreement, then Iran may raise its exports and Europe will be once again able to buy its crude oil, thus leading to a drop in prices.

In general, history is in favor of a reduction in prices of oil and gas, since any previous period of high prices and tight production was followed by a period of oversupply and lower prices.

Based on all the above, any normalization can be expected a few years down the road and it looks like 2023 will be another hard year for Europe, if not harder than 2022. The bet is to maintain natural gas storage at adequate levels, at least equal to last year's, or even higher in order to avert a domino effect leading to the next winter. This will allow for a gradual improvement in conjunction with the rest of the variables.

However, latest estimations are not positive concerning gas storage, since most analysts believe it is going to fall lower than last year at the end of March 2023.

The weather will also play a crucial part, not just for the winter of 2022-2023, but looking forward. On one hand, Europe does not want extreme cold or heat that would increase energy demand, but it also wants to avoid extreme weather effects, such as the drought of 2022 that limited hydro availability and its nuclear production. We should keep in mind that a colder winter than usual can increase natural gas demand by 5%. During a time when Europe hunts for every available molecule of gas in the market, this is a significant percentage.

Realistically speaking, it will be very hard for prices to return to 2019 levels without Russian gas. However, there is the possibility of reduced prices compared to very high current levels, a fact that would bring economic benefits and a kind of relief.

B) Energy-sacrifices on the crisis altar

A second scenario that gains ground based on developments is that the result of the crisis will not be a return to the good old times, but a rearrangement of the European energy sector with unforeseen and mixed consequences.

On one hand, Europe ends the use of Russian gas for the foreseeable future, but it replaces one addiction with another, namely LNG. Since LNG is a product with pronounced volatility, it may not be the right solution in the long term, especially based on prices.

At the same time, the EU's energy taxonomy sets its own limitations and pushes for renewable gasses, therefore LNG will have to be replaced eventually and time is against it.

A second energy-sacrifice with unforeseen repercussions is the changing of the gas benchmark from TTF to a new contract preferred by Brussels and member-states. The EU has said that it aims to introduce a contract based on LNG prices in Europe's terminals. However, there is the issue of liquidity, since such a new market will constitute of fewer players compared to TTF and it will take time for the new contract to gain the companies' trust. In short, the contract will have to gain its commercial worthiness and popularity in practice, something not possible simply through a political decision.

Another energy-sacrifice with complex effects on prices and supply is the grand bet of electrification that calls for the replacing of fuels with electricity wherever possible technologically and commercially. Electric vehicles and other replacements gradually increase power demand and this effect will become more pronounced at the end of this decade. Many European countries have pledged even to end internal combustion engines after 2030, therefore the change will become sharp after a certain point. The electric system and the grid will be hard pressed to cover this higher demand and handle new renewables installations at the same time. Also, there will be increased need for power production, while oil companies are going to lose a part of the demand for their products.

But perhaps the biggest energy-sacrifice of all is the one that concerns energy and climate goals for 2030, that call for renewables penetration at 45% in electricity and a reduction of CO_2 emissions by 55% compared to 1990.

While the use of lignite and hard coal is increased in Europe through the reactivation of plants and mines, people turn to wood for heating and there is more short term laxity towards other solutions with high CO2 emissions, one can see that they cannot be continued indefinitely because climate goals will be threatened.

Obviously, this is not just a matter for Europe, but the planet as a whole. According to latest information, China aims to add more coal plants than previously in order to support its own energy production and reduce gas imports. The country plans 270 GW of new thermal capacity during the next five years, according to China Energy Engineering Corp., while other estimations mention 100-200 GW, a very high number regardless, which will lock high emissions for decades to come. According to Shell's CEO, Ben van Beurden, China increased its coal consumption during a single trimester in 2022 more than the entire energy output of his company.

Therefore, it is evident that the trilemma between security-prices-sustainability has shifted decisively in favor of the first and against the third. It remains to be seen how sustainability will be affected by this situation and if what is temporary becomes permanent.

Another aspect of Europe's current predicament that will determine things for 2023 as well, is gas storage. Since experts mostly agree that even under normal weather conditions, storage will fall below 2022's 30% level on March 2023, the effort to replenish it will be even more difficult. Until August 2022, European storage was filled with the help of still flowing Russian gas, but in 2023 it must be filled from the beginning without most of it. If storage levels are low in October 2023, then the following winter may be even harder than this year's.

Last but not least, there is one more energy-sacrifice that has to do with markets and their operation. The European market became so regulated in 2022 that it would be a joke to say that this is a free market. Price ceilings were set, companies were bailed out or nationalized, but there were also plenty of interventions and exceptions that constrict the market's free operation as it was planned. Are we witnessing a return to an older and more centrally planned market course?

According to Foreign Affairs[28]:

"In addition to economic nationalism and deglobalization, the coming energy order will be defined by something that few analysts have fully appreciated: government intervention in the energy sector on a scale not seen in recent memory. After four decades during which they generally sought to curb their activity in energy markets, Western governments are now recognizing the need to play a more expansive role in everything from building (and retiring) fossil fuel infrastructure to influencing where private companies buy and sell energy to limiting emissions through carbon pricing, subsidies, mandates, and standards."

This new state-ism is expected to become a major issue in each country's upcoming elections.

C) "Lehman Brothers" contagion and an economic crash

Previously we analyzed the mechanism through which the energy crisis may lead suppliers to bankruptcies across Europe. Governments and the EU Commission itself have pledged to support these companies with many billions in order to avert their exit from the market that would lead to dramatic consequences with chain debt reactions, concentration and reduction of competition, as well as consumers being on the edge and general panic.

Given the complicated relations in today's energy trade, with long term contracts of many kinds, interdependence of producers and industries, wide participation in energy exchanges and a multitude of claims, one can see that margins are short and to a certain extent already exhausted.

We only have to remember what happened in the U.S. in 2008 and how the crisis was accelerated as a result of CDOs that steeply multiplied the effects. In this case, the fear is that state guarantees will not be enough if big market players are threatened and they could carry the entire economy in a downward spiral with debt detonations.

Another aspect is that intense energy shortages would increase the chances for states and companies adopting egotistical behaviors, something that would undermine European solidarity, the foundation of the EU's plan to handle the crisis. Greece could cut gas supplies to Bulgaria, Italy could withhold the gas that Greece has stored in its facilities and so on. Indeed, in such a case each country could even use fake technical issues as an excuse, such as the ones Russia claimed for Nord Stream 1.

There are already signs that this is taking place. One example is the power interconnection between Greece and Italy that breaks down in periods of high prices in order to maximize profits for Italian producers. Also, France is rumored to examine the possibility of restricting electricity exports to Italy for a period of two years, while the German operator said that export restrictions may be needed in the future.

Solidarity is now a pressing issue, since it can have decisive results for the course of the crisis. Since the European system is interconnected, there are many interdependencies affecting member-states. If Bulgaria, for example, stops receiving natural gas through the Greek LNG terminal, it may once again turn to Gazprom for renegotiation. This would be a big hit for European energy and foreign policy, so there is a very subtle political balance involved.

But lets return from the geopolitical part to the economic and energy effects. A possible market crash would automatically derail REPower EU and other European efforts by reducing financing capabilities and increasing already high needs for supporting the population, as huge public and private funds would be under threat. At that point, the attention would turn from energy to the wider public and private debt.

We should keep in mind that high subsidies against other budget priorities, billions towards problematic energy companies, the acceleration of new energy projects through public funding and all other special needs as a result of the crisis inexorably lead to an increase of debt.

Indeed, this new reality comes during a period when very low interest rates of past years flooded the economy with money and increased inflation. The effort by central banks to restore the balance by rapidly increasing rates, may hit the obstacle of increased needs for new money in order to face the crisis. Therefore, it is entirely possible that the "money printer" may be turned on once again soon with unpredictable consequences.

Finally, there is also the political aspect. Since European governments are more unstable today compared to the past and are under threat by the high political cost of expensive energy, an economic crash or massive power shortages could lead many countries to premature elections. So far, the energy crisis has already led to limited demonstrations and public discontent, but also to some changes in government, such as in Italy and Bulgaria. In the U.K. there is already a movement for not paying the energy bills and wider economic hardship would cause a multiplication of outrage and distrust on behalf of citizens.

Apart from all the above, we could also add special circumstances that could make the crisis even harder. For example, a typhoon in the Gulf of Mexico could shut down LNG plants for a while with the result of Europe losing part of its much needed supplies. Additionally, one cannot exclude further sources of geopolitical upheaval in places like the Middle East.

In any case, we should not forget the "elephant in the room" that is China, the largest energy consumer globally, whose economy has not yet returned to levels of activity as before the pandemic as a result of continued lockdowns. Until today, the under-performance of the Chinese economy was a major limiting factor for energy demand and prices. If president Xi Jinping manages to reaffirm his power at the end of 2022, it is possible that he will choose to lift lockdowns, thus increasing energy demand and making things harder for Europe.

Finally, another important topic is the effects of the energy crisis for food. Since energy prices directly affect food production globally, their rise could even lead to uprisings in developing countries, like the ones that happened during the Arab Spring or more lately in countries like Sri Lanka.

According to Fertilizers Europe, 70% of European fertilizer production capacity had been curtailed by the end of August 2022. The steep reduction in fertilizer output puts additional pressure on food supply chains.

7. Our energy future

The energy crisis has already led to huge developments, especially in Europe, which affect the whole of economic activity and the very relationship between the average citizen and energy.

In order to find out whether the crisis will have long reaching effects, it matters if today's short term decisions can become permanent in the end.

A fundamental point showing this is the percentage of Russian gas that will be replaced with new renewables and other technologies. This turn is permanent and probably irreversible, so these green projects are not expected to be replaced with Russian gas again, even if there is some kind of rapprochement between the EU and Moscow. Therefore, the crisis leads directly to an acceleration of renewables and developing new technologies, such as green hydrogen and storage.

Additionally, natural gas becomes less competitive for an unknown period, since it is very expensive, therefore Europe's struggle to replace it can become a permanent policy, especially if seen under the light of climate change. This is confirmed so far by REPower EU.

Most experts and analysts agree that high gas and power prices will accompany us for a few years to come, meaning that they will remain higher than before the crisis, although they could be reduced compared to the very high levels we witnessed in 2022.

Klaus-Dieter Maubach, the CEO of German Uniper, said that "prices will not remain at the levels we saw lately, but we believe that they will not return to previous levels."

As a result, natural gas loses its competitiveness, at least for the mid-term, against other sources, especially renewables.

A second fundamental question has to do with coal plants that were called upon in 2022 in many countries to replace natural gas and maintain security of supply. CO_2 emissions of states that used coal are expected to rise considerably and it is evident that they cannot be allowed to do so indefinitely, since it would threaten the climate effort.

On the other hand, Brussels have cultivated all this time a mood of understanding and sent the message that climate protection can be downgraded for as long as the crisis rages, in favor of security. Based on the above, it would be hard to say when these coal plants will shut down again and what it will mean for European climate efforts. Similar developments may take place in other economic sectors. Already, the fuels lobby has asked the EU to postpone the date of banning internal combustion engines, scheduled for 2035, as a result of extraordinary circumstances created by the crisis.

Another "safe" bet is that the energy saving effort will gain more traction from now on. Since on a European level there are now measures that connect the subsidy level for consumers to their level of energy savings, it is evident that Brussels finally recognize the crucial importance of energy efficiency, even while forced by developments.

Perhaps the crisis will lead to a permanent energy saving culture for the citizens themselves. Their consciousness changes because of the need and high cost and they will turn more and more towards solutions such as house insulation, more effective heating etc.

As we examined, there are also some positive factors that may soften the blow from now on. One of them is France, where during the winter many nuclear reactors are expected to return to operation.

Other extraordinary factors, such as drought, are expected to cease and lead to more normal conditions, although extreme weather phenomena of other types cannot be excluded given the effects of climate change.

113

A) A new course for the European market

During a previous chapter we analyzed the operation of the European power market, where TTF has a major role. These days in Brussels there is a wide conversation about how to change the architecture of the market, not just for the duration of the crisis, but permanently, in order to better reflect fundamental forces and lead to fair prices for consumers.

The EU Commission's president, Ursula von der Leyen, admitted openly that the current model is problematic and called for its reform. "Consumers must enjoy the benefits from cheap renewables. This is why we must disconnect the dominant price of natural gas from the electricity price," she said.

Now, experts working for the EU Commission together with member states and the energy sector are engaging in a long dialogue that may lead one day to a new market model.

We should take into account that different interests between member-states, various energy technologies and businesses will fight hard to protect themselves and acquire long term benefits, since the new model will define market operation possibly for decades to come.

So far, many proposals have been submitted and are examined, but the basic idea is to replace the marginal price in the energy exchange with a calculation that will reduce the wholesale price formation from the price of natural gas.

Experts who stand in favor of changing the market suggest that two different wholesale markets could be formed, one with natural gas plants and one with renewables and other less expensive sources. The final wholesale price will be formed by both and not be affected that much by TTF.

Another proposal is not to change the fundamental nature of the existing model, but improve it through stricter oversight. As part of this idea, offers made by power producers will be under higher scrutiny than before, exactly as it happens to many markets in the U.S. The goal is for natural gas plants to present offers based on their true operational cost, not just the TTF price.

However, Jérôme à Paris warns:

"Expensive marginal production is suddenly called a lot more often - and has been made more expensive by higher gas prices. But there is no substitute to flexible gas-fired plants that are usually idle - all the cheaper capacity is already producing... If you cut the price link the gas producers won't join the market and you'll have brownouts."[29]

Gradually over Europe a consensus is formed for changing the model, although it is unknown which direction it will take and how deep the changes will be. The new model must be able to produce fair prices, to support the green transition and to cover different interests of member-states. Therefore, this process is expected to last long.

Furthermore, Europe also discusses a price cap to gas imports, as well as replacing TTF with a new benchmark. These are all very difficult issues from a political, regulatory and legal perspective and they increase complexity and the degree of difficulty for reaching commonly acceptable decisions.

B) What is the ceiling for renewables?

The boost for renewables is now a given, especially in Europe, since the EU, the governments and most citizens are convinced that they are the only way forward. Indeed, renewables supporters believe that as we leave behind the era of expensive renewables, their future growth will be easier. We already made the first hard steps and the next ones will be smoother and cheaper.

Regardless, renewables cannot replace conventional production in the primary mix on their own, not even in the electricity mix. They also need investment in technologies such as storage (batteries, pump hydro etc.) as well as green hydrogen in order to fully take over the weight of fossil fuels in power production and fuels and lead to decommissioned conventional plants, but also changes in transportation, industry and heating/cooling.

Since these two technologies are still at an intermediate level of development, their cost remains high despite the progress made during previous years.

According to the EU Commission, the cost of producing green hydrogen was 3-6,55 dollars/kg in 2020, while hydrogen from "dirty" sources cost about 1.80 dollars and "blue" hydrogen made by natural gas cost 2,40 dollars. The set goal is for green hydrogen to reach 1-2 dollars in order to be directly competitive, meaning a third compared to today[30].

In the case of lithium batteries, who are the most common energy storage technology, the cost was reduced from 1,200 dollars/KWh in 2010 to 132 dollars in 2021. The problem is that price rises in raw materials pushed up the cost of batteries during the last two years by at least 30%, therefore the crisis affected them negatively. Meanwhile, at the end of 2021, 20.84 GW of battery storage was operational worldwide and is expected to reach 92.22 GW by 2026, according to GlobalData.

American MIT calculated in 2019 that battery cost must be further reduced by 90% to 20 dollars/KWh in order to be able to support a 100% renewables penetration in the future, which is the desired result of the energy transition[31].

One troubling part of the equation is that during the first half of 2022, prices of raw materials skyrocketed and then retreated, with the single exception of lithium, which remained high.

Based on the above, when we consider the cost of the future green energy system, we much take into account not just the low cost of renewables themselves, but also the higher cost of storage and green fuels in order to have a complete picture. If we add on top the even higher cost of other immature green technologies, such as solar windows, tidal energy, gravitational storage etc. then the curve becomes even steeper.

Unfortunately, there is no guarantee that the course of storage and hydrogen cost will be the same as it was for wind and solar farms in the past. It is a bet than Europe and the planet must win through tough efforts and constant investment that become harder during an environment of high equipment cost. Whether this cost reduction occurs or not is one of the fundamental aspects for the course and the success of the energy transition.

Accordingly, another critical issue is to find and exploit the necessary materials for new energy technologies. According to the World Bank, "the adoption of low carbon power production translates to a permanent increase of demand for copper, nickel, cobalt and lithium."[32]

McKinsey estimates that lithium demand, an element used in batteries, is going to increase globally from 500,000 tonnes in 2021 to 3-4 million tonnes by 2030. IEA expects that by 2040 lithium needs will become 13-

51 times higher than 2020, as batteries use will be expanded. These are obviously enormous quantities[33].

At the same time, lithium prices have skyrocketed and in the fall of 2022 they reached 71,000 dollars/tonne in China, three times higher than a year earlier.

Refinitiv believes that "we simply do not have enough metals and minerals to cover the insatiable and rising demand and achieve goals set by the Paris Accord and COP26."[34]

Lithium is not the only element where supply is an issue. The green transition is expected to raise demand for so-called rare earths, a group of 17 chemical elements with exotic names, such as neodymium and dysprosium, which are necessary for a series of new technologies like wind turbines and electric vehicles. IRENA estimates that by 2030 demand for permanent magnets will increase from 50,000 to 225,000 tonnes annually, with 180,000 tonnes required for electric vehicles and 50,000 for wind turbines[35].

When it comes to Europe, it is expected to need 35 times more lithium by 2050, as well as 7-26 times more rare earths, according to calculations by the Belgian Katholieke Universiteit. Furthermore, the continent will require 33% more aluminum, 35% more copper, 45% more silicon, 100% more nickel and 330% more cobalt[36].

Geopolitical competitions and trade protectionism may turn the effort to acquire these materials even harder, leading to the rising of new "Saudi Arabias" that will control supply as monopolies, but also trade or even real wars for control. At this point, we should remember Elon Musk's statement from 2020, concerning Bolivia and restricted lithium supplies: "We will coup whoever we want! Deal with it."

The conclusion is that there may be insatiable demand for growing green energy globally, but it is still uncertain whether it can be covered in terms of materials and cost reduction.

The fact that the global economy is presently going through a phase of reverse globalization may lead to more trade restrictions and fewer possibilities of acquiring necessary raw materials for Europe in the future, especially if rivalries and geopolitical spats increase.

We should also consider Vaclav Smil's words on the subject:

"The persistence of new energy myths is at least as remarkable. New energies are initially seen to carry few, if any, problems. They promise abundant and cheap supply, opening up the possibility of near-utopian social change. After millennia of reliance on biomass fuels, many nineteenth-century writers saw coal as an ideal energy source and the steam engine as a nearly miraculous prime mover. Heavy air pollution, land destruction, health hazards, mining accidents, and the need to turn to progressively poorer or deeper coal reserves soon swept that myth away. Electricity was the next carrier of unbounded possibilities, its powers eventually so far-reaching that they would cure poverty and disease.

What can be foreseen with great certainty is that much more energy will be needed during the coming generations to extend decent life to the majority of a still growing global population whose access to energy is well below the minima compatible with a decent quality of life. This may seem to be an overwhelming, even impossible, task. The global high-energy civilization already suffers economically and socially from its precipitous expansion, and its further growth threatens the biospheric integrity on which its very survival depends."[37]

In the end, extraction, production and transportation of materials and equipment for green technologies will continue to depend for many years on the use of fossil fuels used in machinery, factories, ships, trucks etc. If investment on fossil fuels is reduced and demand remains high, then their price will rise, affecting the cost of green technologies that we need to replace them. It is a viscous circle with no easy solution.

C) *The hard riddle of energy demand and growth*

As projections about the future growth of green electricity and fuels are uncertain based on current trends, demand must be addressed.

If energy demand was lower, then it would be easier to cover it through green technologies, while decarbonization would advance much faster. As we mentioned at the beginning of the book, the size of the energy "pie" is what sets to a large degree the level of difficulty and the success of the energy transition.

Demand is an issue closely related to economic growth, since historically the two are highly connected. This fact is understood no just economically, but also through biology and physics: When a colony of organisms gains access to bigger energy resources, then it grows rapidly and on the contrary, when it loses that access, it shrinks. This is a global constant that should trouble us given the speed at which the global economy grew in the recent past, but also the direction it is taking today.

There are several schools of thought globally about this issue. A traditional liberal economist would tell us that the crisis is the logical result of the rearrangement taking place in the energy sphere and the laws of the market will lead to a new point of balance in time. As growth will be further disconnected from energy consumption, it may continue unhindered. If this is true, then his/her view is probably right.

On the other hand, more radical thinkers claim that our technical civilization has reached its material and ecological limits and what we are witnessing these days is the natural outcome of this fundamental event. They believe that today we are experiencing the "revenge" of externatilities, which means that for more than 100 years we ignored the role of environmental degradation, we misspent fossil fuels, while the ability to exploit the natural environment for human needs has begun to erode, as well as natural services.

If this is true, then energy demand and economic growth will start to reach a plateau and then they will recede. The issue is whether this will take place in an organized way through wise political choices or chaotically through crises.

Concerning this issue, writer and economic anthropologist, Jason Hickel, says:

"Imagine next year we cut fossil fuel use by 10%. And then the following year we cut it by another 10%. And so on the next year and the next. Even if we throw everything we have at building our renewable energy capacity and improving energy efficiency—which we must do as a matter of urgency—there's no way we can cover the full gap. The truth is that rich countries are going to have to get by with less energy. A lot less.

How can we possibly manage such a scenario? Well, in the existing economy it would be sheer chaos. The price of energy would skyrocket. People would be unable to afford essential goods. Businesses would collapse. Unemployment would rise. Capitalism—which depends on perpetual growth just to stay afloat—is structurally incapable of sustaining such a transition."[38]

At the end, the issue of energy in its entirety must be viewed under the prism of what constitutes progress. There is no doubt that the rise of energy consumption was a primary factor in the 20th century that led to a rise of quality of life for most people on the planet, with an improvement in life expectancy, education, freedom of movement, better communication and many other freedoms and comforts.

But there are also signs that after a certain point, economic growth and higher energy demand no longer bring proportionally positive advantages and progress. At the same time, environmental destruction and climate change are worsened through humanity's ever increasing appetite for resources and energy of all kinds and finally lead to a degradation of human life quality.

There comes a point that for example, to drive an SUV or buy appliances that are made to last only a few years is not worth it if we could see the environmental and economic destruction they are causing in the future on a collective and therefore on a personal level. Future cost is largely hidden from the price of the products, since it does not include externalities. In essence, we are paying for the products today, but we neglect that we are going to pay for them again tomorrow.

The added benefit from these pleasures is often dramatically lower than their full cost. Perhaps this should be the very message of the energy crisis we are going through: That there is plenty of energy "fat" that humanity can cut without losing the important elements that constitute progress, such as health services, scientific discoveries, good nutrition and many others.

In 2020, IEA calculated that increased electric vehicle sales led to a saving of 40,000 barrels of oil daily in the world. However, increased SUV sales completely canceled this benefit, since these vehicles consume

20% more energy per kilometer than a vehicle of average size.

The same is true for vehicle emissions: Electric vehicles and fuel consumption improvements reduced CO_2 emissions by 350 megatons from 2010 to 2020, but SUV emissions rose by over 500 megatons during the same period.

On another level, Nate Hagens put it like this:

"Every day we are transforming millions of barrels of oil into micro-liters of dopamine in order to feel momentary enjoyment through the 'likes' on social media and the other things that grab our attention and we consider immaterial, but they require energy and resources just like everything else."

A part of energy and environmental experts who turn against growth as an index of progress, believe than deep down the nature of the issue is not even about energy or the climate. Even if tomorrow we could have a completely clean, cheap and reliable energy source, or if we could suck excess CO_2 from the atmosphere, humanity's appetite for expansion against the natural environment would continue because this is our economic model. In such a case, we would have sold climate change, but all the other environmental problems would remain. The limits set by natural resources and natural services availability would not change.

This is why a future of 100% renewables does not automatically mean solving the environmental issue, but only one part of it. This is why the energy issue is at its core moral and not economic or technological.

Demand reduction for energy and resources must become the primary goal for the future and through it all side-goals can be achieved, if we want to succeed in the energy transition.

Perhaps the right recipe for our future contains things like standards for long lasting appliances and the right to repair, wide energy saving programs for building insulation, a rapid expansion of public transportation, as well as renewables.

8) Epilogue

The energy crisis was the result of a plethora of factors, structural and extraordinary, which coincided and led to profound situations. However, it also reflects more fundamental choices of our society and greatly concerns the priorities we set as a civilization, therefore it is not merely a technical issue for experts to solve, or politicians, economists, engineers and diplomats to handle.

Whether we are talking about sustainable development that is the hope of the traditional elites, or some kind of return to a more balanced relationship with nature and energy that new voices call for, it certainly will not happen from the top, exclusively through political programs, investment plans and international deals. It requires a change in attitude for citizens about what energy means and a redefinition of our relation to them both quantitatively, as well as qualitatively.

It is the very decentralized nature of new energy technologies that demands their choosing and application on behalf of citizens, since we are talking about photovoltaics on homes, apps that show energy consumption in real time, smart devices, micromobility and other ideas that are related to personal choices and knowledge. At the end, the citizens will determine our energy future through their behavior and the political choices they are going to support.

During a crisis, the temptation to be shortsighted and choose the immediate profit, the lowest damage or how to make it through the winter or the next few years is high. However, history has shown that states and societies that were able to resist it and look further will be rewarded and will probably become leaders in the new landscape.

What the crisis has shown is that what is at stake for Europe is not just the climate effort and sustainability, but also long term energy security and industrial competitiveness. Demand reduction has the ability to help in all these issues, leading to a "leaner and meaner" Europe in the future. Perhaps one day, other parts of the world, like the U.S. will be faced with similar crises, while the EU will have become greener, more resilient and self-reliant.

In order to have this great future and avoid further crises, difficult decisions will have to be made on behalf of governments and citizens around Europe. This continent has shown its resiliency in the past and must do so again now.

Bibliography – References

1. 2021, BP Statistical Review 2022

2. Griffiths, Sarah. (6/3/2020). "Why your internet habits are not as clean as you think", BBC

3. Smil, Vaclav. (2022). How the world really works. Viking

4. McGreal, Chris. (30/6/2021). "Big oil and gas kept a dirty secret for decades. Now they may pay the price", The Guardian

5. International Renewable Energy Agency – IRENA

6. Mitchell, Russ. (5/9/2022). "The energy historian who says rapid decarbonization is a fantasy", LA Times

7. IEA, Tracking Clean Energy Progress, 2022

8. Jerome a Paris, "How can we discuss the energy crisis when journalists are so ignorant?", Substack

9. Heron, James. Smith, Grant. Crowley, Kevin. Hurst, Laura.

(16/2/2022). "Oil's Spectacular Covid Crash Set the World Up for $100 Crude", Bloomberg

10. IEA, World Energy Investment, 2022

11. U.S. Energy Information Administration – EIA

12. Intercontinental Exchange

13. Eurelectric, Power Barometer, September 2022

14. Jacobs, Justin. (29/8/2022). "ExxonMobil and Chevron shatter profit records after global oil price surge", Financial Times

15. IEA, Renewable Energy Market Update - May 2022

16. Simon, Frederic. (9/9/2022). "Power firms warn about 'unprecedented' liquidity crisis in Europe", Euractiv

17. Shiryaevskaya, Anna. (6/9/2022). "Energy Trading Stressed by Margin Calls of $1.5 Trillion", Bloomberg

18. Thomas, Nathalie. Sheppard, David. Milne, Richard. (5/9/2022). "'Lehman Brothers moment': EU's energy market crisis

rises", Financial Review

19. Thomas, Nathalie. Stafford, Philip. Sheppard, David., (5/9/2022). "Power producers call for collateral change to avert 'Lehman' moment", Financial Times

20. Carbon Tracker: "Global Gas Power", 2021

21. Gimber, Hugh. (September 2022). Exploring the economics of Europe's energy crisis, JP Morgan

22. World Economic Forum, "What are European countries doing to reduce the impact of rising energy prices on homes and businesses?", September 2022

23. Brusuelas, Joseph. Pugh, Thomas. (September 2022). "European energy crisis: A quandary with no good policy options", The Real Economy Blog

24. EuroCommerce, Retail and wholesale and the energy crisis – an urgent need for support, September 2022

25. Alderman, Liz. (19/9/2022), 'Crippling' Energy Bills Force

Europe's Factories to Go Dark", New York Times

26. Eurometaux, "Europe's non-ferrous metals producers call for emergency EU action to prevent permanent deindustrialisation from spiralling electricity and gas prices"

27. Rystad Energy, "Spurred by the energy crisis, global LNG investments will now peak at $42 billion in 2024, a 50% jump from current spending", 24/8/2022

28. Bordoff, Jason. L. O'Sullivan, Meghan. (August 2022). "The new energy order", Foreign Affairs

29. Jérôme à Paris, "Reminder: the link between gas and power prices makes sense, today", Substack

30. European Commission, Hydrogen Strategy, 2020

31. MIT, Storage Requirements and Costs of Shaping Renewable Energy Toward Grid Decarbonization, 2019

32. McKinsey, "Lithium mining: How new production technologies

could fuel the global EV revolution", April 2022

33. IEA, "Mineral requirements for clean energy transitions"

34. Refinitiv, "Metals and minerals, the blind spot of the energy transition", 2022

35. IRENA, "Critical materials for the energy transition: Rare earth elements", 2022

36. Eurometaux, Katholieke Universiteit. "Metals for Clean Energy: Pathways to solving Europe's raw materials challenge", April 2022

37. Smil, Vaclav. (2017). Energy and Civilization: A History. Page 437. The MIT Press

38. Hickel, Jason. (November 2021). "What Would It Look Like If We Treated Climate Change as an Actual Emergency?", Current Affairs

About the author

Charalampos Aposporis is an energy journalist from Athens. He has studied International and European Relations with an MA degree on Security and War. He is currently working for energypress.gr/energypress.eu, the leading energy news portal in Greece. He is married with two daughters.

Printed in Great Britain
by Amazon